Multimedia MathPro 4.0

Student Version

Fifth Edition

Elementary
Algebra for College Students

Allen R. Angel

PRENTICE HALL
Upper Saddle River, NJ 07458

Media Editor: Audra Walsh
Special Projects Manager: Barbara A. Murray
Production Editor: Ashley Scattergood
Supplement Cover Manager: Paul Gourhan
Supplement Cover Designer: PM Workshop Inc.
Media Buyer: Tom Mangan
Cover Photo: James H. Carmichael, Jr./ The Image Bank/Lyre
        Formed Volutes "Lyre Lyraeforms" E. Africa

Printed in the United States of America

10 9 8 7 6 5 4

ISBN 0-13-040291-5

Prentice-Hall International (UK) Limited, London
Prentice-Hall of Australia Pty. Limited, Sydney
Prentice-Hall Canada, Inc., Toronto
Prentice-Hall Hispanoamericana, S.A., Mexico
Prentice-Hall of India Private Limited, New Delhi
Prentice-Hall (Singapore) Pte. Ltd.
Prentice-Hall of Japan, Inc., Tokyo
Editora Prentice-Hall do Brazil, Ltda., Rio de Janeiro

Contents

Welcome to Multimedia MathPro Explorer 4.0!

MathPro Explorer is a computerized version of your math textbook and is organized in an identical format. For each section and objective in your text, there is a corresponding section in MathPro Explorer.

MathPro Explorer generates a set of problems to solve for each learning objective. Unlike textbooks, which only show a few examples in each section, MathPro Explorer allows features that a textbook cannot:

1) View multiple examples of similar problems.
2) View the complete solution to any problem.
3) Work through an interactive step by step process on how to solve a particular problem.

MathPro Explorer can be purchased in two versions:

1) The School/Network Version is available for installation in multi-user environments either via a LAN or installed on shared lab computers.

2) The Home/Student Version is available for installation on an individual computer for an individual user.

Details about these versions and possible upgrade paths for the versions are described below.

School/Network Version

The School/Network Version of MathPro Explorer provides the complete functionality of the MathPro Explorer Tutorial product. This includes the following, which are structured according to the Chapter, Section, and Objective content of the textbook:

- Problem Generation
- View and/or Watch Example
- Step by Step
- View Solution
- Redo Problem
- Check Answer
- Score Summary

The School/Network Version also provides the capability to generate and take Practice Tests and receive and reply to messages from the administrator.

An additional Exploratory element, the "Explorations Rights" upgrade option, is locked and hidden within the product and mentioned in the Introduction Video. The video is available from the extended Welcome Back screen and the Course Introduction button on the Chapter screen. The Explorations Rights can be enabled for an individual by following the "Key Disk" procedure detailed later in this document.

In conjunction with MathPro Explorer, a separate Administrator Program allows for instructors to perform administrative functions.

Home/Student Version

The Home/Student Version of MathPro Explorer provides the complete functionality of the MathPro Explorer Tutorial product mentioned above, as well as an Exploratory element. The Explorations element adds Exploratory Labs that employ the following additional tools, each with extensive capabilities:

- Algebra Tools
  ⇒ Symbolic Editor
  ⇒ Assignment Editor
  ⇒ Algebra Tiles - Manipulative
- Coordinate Graph Tool
- Geometry Tool
- Spreadsheet Tool
- Data Series Graphs
  ⇒ Bar Graph
  ⇒ Line Graph
  ⇒ Pie Graph
  ⇒ Histogram Graph
  ⇒ Box and Whisker Graph
- Probability Tools
  ⇒ Spinner
  ⇒ Number Objects
- Text Tool

Unlike the School/Network Version, Exploration Rights are enabled automatically the first time the user logs in to MathPro Explorer with his/her MathPro Explorer username. On subsequent logins, the program will automatically log the student in using the previously entered username and display the Welcome Back screen. On each subsequent login the user will also be given the opportunity to create a personal Key Disk which can be used to enable Exploration Rights for that single individual on school computers running the School/Network Version. (The ability to create a Key Disk is also available from the Summary screen in the program.)

NOTE: The Key Disk will automatically use the first and last login names from the Student Version, thus, be aware that the first time that you log in at home you must enter the same username that you use at school.

Access Rights Key Disk

User access rights via a Key Disk control access to the Explorations. You may purchase a Key Disk as a separate upgrade product or create a Key Disk from the Home/Student Version of MathPro Explorer. The Key Disk diskette will contain the specific access rights purchased. You only have to run the Key Disk program once per computer with your MathPro Explorer username to establish and maintain your upgrade access rights.

Key Disk Upgrade Procedure

A Key Disk is needed to enable Exploration Rights for the School/Network Version of MathPro Explorer. (As mentioned above, these Rights are automatically enabled in the Home/Student Version.)

To Create a Key Disk:

1) From the Chapter screen, click the Summary button.
2) In the bottom right corner of the screen, click the Create Key Disk button.
3) Insert a diskette in the diskette drive and click OK.

You may bring the Key Disk to the computer lab at school to access Explorations.

To Execute the Upgrade:

Note: In the following instructions, username refers to the name that you use to log in to MathPro Explorer.

1) Exit MathPro Explorer if MathPro Explorer is running.
2) Insert the Key Disk in the diskette drive.
3) From File Manager (Win 3.1) or Windows Explorer (Win 95) select the diskette drive (typically a:) to see the contents. On the Macintosh, double-click the diskette icon displayed on the desktop to see the contents. There are three files on the Key Disk, one of which is an executable file.
4) Double-click the KeydskEX.exe file to install Exploration Rights on the computer.
5) Enter your MathPro Explorer username and Click OK. Remember to use your MathPro Explorer username. (Note: If you create your Key Disk from the Home/Student Version, your username will be entered for you by the program and will be protected. This ensures that the Key Disk enables your rights at school and not someone else's.)
6) Log in to MathPro Explorer with your username. You will have Explorations Rights enabled. The Explorations and Toolkit buttons will now be active and visible at the bottom of the screen, and the Explore button will be displayed on the Problem Sets Screen. See the Explorer Upgrade sections in this document for details.

The Key Disk is specific for a single person on a single computer. A user returning to the same computer will not need the Key Disk again to access Exploration Rights. If a user moves to another computer on subsequent visits to the lab, the Key Disk initialization procedure must be repeated.

Multimedia MathPro Explorer v 4.0 System Requirements

WINDOWS REQUIREMENTS

- 80486/66 MHz processor minimum
- 12MB of random access memory (RAM);
- 16MB or more recommended
- 30MB free hard disk space or network drive space for program
- Microsoft Windows 3.1, Windows 95 or higher
- CD-ROM drive
- A graphics adapter card (VGA, Super VGA, or other Windows-compatible card) capable of displaying at least 256 colors at 640 X 480 pixel resolution
- Sound card
- Quick Time 3.0

MACINTOSH REQUIREMENTS

- Power Macintosh or higher
- 12MB of random access memory (RAM);
- 16MB or more recommended
- 45MB free hard disk space or network drive space for program
- Macintosh System 7.5 with Virtual Memory 'On'
- CD-ROM drive
- Color Display
- Quick Time 3.0

Windows Home/Student Install Instructions:

1. To install MathPro Explorer 4.0, run the setup.exe on the installation CD and follow the directions on screen.
2. For Windows 3.1, the setup will check to ensure you are running the correct version of Win32s. If you do not have Win32s or are running an older version you will not be able to run MathPro Explorer 4.0. You will be prompted to confirm installation for Win32s or to Exit Setup. Chose 'Yes' to continue and install the correct version of Win32s.
3. Choose a destination location. To install to a directory other than the default, click the Browse button. MathPro Explorer 4.0 can only be installed to a drive on your computer, typically "C:\".

   IMPORTANT: It is strongly recommended that you install all MathPro Explorer books to the same location. To change your MathPro Explorer 4.0 install location, choose 'Back' and change the Destination Folder.

4. Specify the Program Folder (Windows 95) or Program Group (Windows 3.1).

   IMPORTANT: It is strongly recommended that you do NOT change the Program Folder (Windows 95) or Program Group (Windows 3.1).

5. To confirm selections choose the 'Next' button. To change your selections choose the "Back" button.
6. If you have MathPro Explorer 3.0 books in the same install location, you must run the MPE Converter after you complete the installation of MathPro Explorer 4.0 to upgrade the 3.0 book(s).
7. MathPro Explorer 4.0 requires QuickTime to play introductory (Windows 95 and 3.1) and instructional (Windows 95) videos.

Windows 95
If you do not have QuickTime 3.0 or higher, you must install QuickTime 3.0 after you complete the installation of MathPro Explorer 4.0.

To install QuickTime 3.0 go to: http://quicktime.apple.com/sw/sw3.html

Windows 3.1
If you do not have QuickTime 2.1.2 or higher, you must install QuickTime 2.1.2 after you complete the installation of MathPro Explorer 4.0.

To install QuickTime 2.1.2 go to:
http://www.apple.com/quicktime/download/qtwin3.1.html

8. Once MathPro Explorer 4.0 has been successfully installed, click Finish to complete the installation process.

Upgrading MathPro Explorer 3.0 books
1. Run MPE Converter from the new Program Group (Windows 3.1) or Start Menu-Program Folder (Windows 95).

> IMPORTANT: MathPro Explorer 3.0 books installed in locations other than the current install location will NOT be upgraded to run with MathPro Explorer 4.0.

2. A new Administrator will be created the first time you run MathPro Administrator after the conversion. You will need to recreate the administrator password.

> IMPORTANT NOTE: If you logged in with only a first name (e.g. Jane) in the MathPro Explorer 3.0 book, then the program will use 'Unknown' as the last name (e.g. Jane Unknown) when logging into the MathPro Explorer 4.0 upgraded version. In addition, if you logged in with only a last name (e.g. Smith) in the MathPro Explorer 3.0 book, then the program will use 'Unknown' as the first name (e.g. Unknown Smith) when logging into the MathPro Explorer 4.0 upgraded version. If you logged in as Visitor, the program will use 'MathPro Student' or 'Visitor Unknown' as your first and last names when logging into the MathPro Explorer 4.0 upgraded version.

Access To MathPro Explorer:
Windows 95
You will have a new folder on your local drive called "MathPro Explorer". Each book installed will be within the "MathPro Explorer" directory. Click the Start button and chose Programs to access the Program Folder for the desired book. You may run the MathPro Explorer, the MathPro Help, the Glossary, or the Administrator from this group. Access to the Explorations is available from within MathPro Explorer.

Windows 3.1
You will have a new folder on your local drive called "MPE". Each book installed will be within the "MPE" directory. From Program Manager, select Program Group for the desired book. You may run the MathPro Explorer, the MathPro Help, the Glossary, or the Administrator from this group. Access to the Explorations is available from within MathPro Explorer.

Uninstall Instructions:
Windows 95

To uninstall MathPro Explorer 4.0, open the Control Panel and double-click on Add/Remove Programs. Select the appropriate author/title/version of the program to uninstall from the list of installed programs.

Windows 3.1
To uninstall MathPro Explorer, double-click the Uninstall icon in the Program Group.

For any uninstall, the program will check for shared files. If you have other books installed, the program does not remove the shared files because the other books use these files. If there is a question about a file that the program cannot answer, you will be prompted to decide whether to remove the shared file. If the book you are uninstalling is the only one installed, or if you are uninstalling all books, then you can safely remove shared files without risk of deleting files used by other programs. In any case, all files that were created through MathPro Explorer will not be removed. You should check the MathPro Explorer or MPE directory and the book-specific directory and remove any unwanted files and/or the directories.

Macintosh Home/Student Install Instructions

Installing MathPro Explorer 4.0 upgrades all existing MathPro Explorer 3.0 books in the install location.

1. Insert MathPro Explorer Home CD in the CD drive.
2. Double-click the MPE Home Install CD icon to open the folder.
3. Double-click the MPE Home Install icon in the MPE Home Install window. Follow the installation instructions on the screen.
4. Choose an install location. Use the box in the lower left corner to select a different location. DO NOT select a location which contains a period(.) in the path name. For example, *Mac HD:MyNetworkPath:MathPro.Explorer* is NOT a valid path.

   IMPORTANT: MathPro Explorer 3.0 books installed in locations other than the current install location will NOT be upgraded to run with MathPro Explorer 4.0.

   It is strongly recommended that you install all MathPro Explorer books to the same location.

   If you currently have MathPro Explorer 3.0 books in more than one location, then you must run a MathPro Explorer 4.0 Upgrade Install to upgrade 3.0 books that are not in the current install location.

5. Click the Install button in the lower right corner.
6. Click the Quit button when the installation is complete. Any MathPro Explorer 3.0 books in the install location will be upgraded at this time. You will see a message stating that the conversion is complete.

Upgrading MathPro Explorer 3.0 books
1. Any MathPro Explorer 3.0 books in the install location are upgraded by installing MathPro Explorer 4.0.

2. A new Administrator will be created the first time you run MathPro Administrator after the conversion. You will need to recreate the administrator password.

   IMPORTANT NOTE: If you logged in with only a first name (e.g. Jane) in the MathPro Explorer 3.0 book, then the program will use 'Unknown' as the last name (e.g. Jane Unknown) when logging into the MathPro Explorer 4.0

upgraded version. In addition, if you logged in with only a last name (e.g. Smith) in the MathPro Explorer 3.0 book, then the program will use 'Unknown' as the first name (e.g. Unknown Smith) when logging into the MathPro Explorer 4.0 upgraded version. If you logged in as Visitor, the program will use 'MathPro Student' as your first and last names when logging into the MathPro Explorer 4.0 upgraded version.

Accessing the MultiMedia features of MathPro Explorer 4.0
MathPro Explorer 4.0 requires QuickTime to play introductory and instructional videos. If you do not have QuickTime 3.0 or higher on your machine, you must install QuickTime 3.0 after you complete the installation of MathPro Explorer 4.0.

To install QuickTime 3.0 go to: http://quicktime.apple.com/sw/sw3.html

Access To MathPro Explorer:

You will have a new folder on your local drive called "MathPro Explorer Home Version". Each book installed will be within a "MathPro Explorer" directory. You may open (double-click) the MathPro Explorer folder to see the program icons. From here, you may double-click and run the MathPro tutorial program, the MathPro Help, the MathPro Administration, the MathPro Administration Help, or the Glossary. Access to the Explorations, MathPro Explorer Toolkit, and MathPro Explorer Toolkit Help is available from within MathPro.

MathPro Explorer v 4.0 Release Notes

MathPro Explorer 4.0 Multimedia Features

MathPro Explorer 4.0 now offers videos featuring the book's author to go along with selected objectives. These videos are available through the new Media Browser (previously the Explorations Browser) and directly from the problem set screen. Please refer to the following explanations to see how you can take advantage of MathPro Explorer 4.0 Multimedia.

Mac and Windows (except Windows 3.1)

If you are running the **MathPro Explorer Home** version the videos will be run from your "CD Title" CD. Please make sure this CD is loaded when watching videos.

If you are running the **MathPro Explorer School** version the videos will be installed to the book's media directory on the network and copied to the client machine only as needed by the user.   *Windows*
- Videos are copied to the client machine into
  <<Windows Directory>>\mpTemp\<<Book Name>>

  Example:
  C:\Windows\mpTemp\mgintro

- This client directory will not exceed 30M or 20% of free hard drive space.
- If copying a new video will exceed 30M or 20% free hard drive space then the *least* recently accessed video will be deleted from the client machine.

Mac
- Videos are copied to the client machine into
  <<System Folder>>:<<Book Name>>

  Example:
  Mac Hard Drive: System Folder: mgintro

Windows 3.1

MathPro Explorer 4.0 multimedia videos are not available on Windows 3.1. QuickTime 3.0 or higher is required and QuickTime 3.0 is *not* available for Windows 3.1.

If you attempt to run the videos on **Windows 3.1 MathPro Explorer School** version you will get the following message which should be ignored:
"Please contact your system administrator to ensure that the movies are installed properly on your network."

If you attempt to run the videos on **Windows 3.1 MathPro Explorer Home** version you will get the following message which should be ignored:
"Please insert your MathPro Explorer CD."

8

Known Issues

1. MathPro Explorer 4.0 upgrades any existing 3.0 program and books in the same install location. If you install a 3.0 book after installing a MathPro Explorer 4.0, you must run a MathPro Explorer Upgrade Install in that 3.0 directory to ensure that the 3.0 book is upgraded and all shortcut icons work appropriately.

2. When MathPro Explorer is uninstalled, all files that were created by MathPro Explorer will not be removed automatically by the uninstall program. You should check the MathPro Explorer or MPE directory and the book-specific directory to remove any unwanted files and/or directories after the uninstall. Please refer to the MathPro Explorer User Manual for program functionality.

Quick Start

Follow these quick easy steps to immediately start using MathPro Explorer.

Login and click the **OK** button.

Click the **Chapter** you want to work in.

Click the **Section** you want to work in.

Click the **Objective** from which you want to work problems. Warm-up problems or Exercise problems are automatically generated. If you are working Warm-up problems, your answers will not be scored. If you are working Exercises, your answers will be scored.

Click the problem you want to work.

Read the instructions and enter your answer in the space provided. Do not use a period in your answer. For example, "5 inches" can be written as "5 in" but not as "5 in."

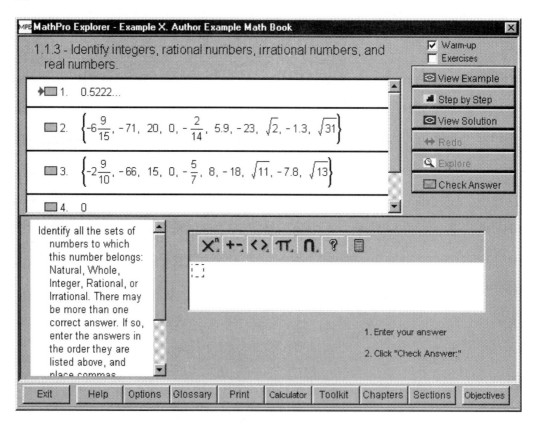

The Login Screen

The Title screen displays a graphic picture of your textbook cover. It will display briefly and be followed by the Login screen.

This screen is where you will log in the first time you run MathPro Explorer. **You will only be allowed to enter your student information once**, so make sure you enter the same username you use at school. On subsequent logins, you will be automatically logged in with you username. This allows MathPro Explorer to track your progress and your results throughout the program.

After you complete your login, you will see a dialog box which will give you the opportunity to play an Introduction Video to MathPro Explorer. From this dialog box you can also launch to the Prentice Hall website for your book. You can choose to suppress this screen for future logins. If you choose to suppress this dialog box, you can access the Introduction Video and the launch to the website from the Chapter screen.

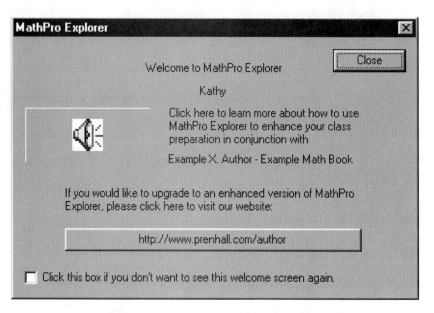

If you have previously logged in, and if you have suppressed the default Welcome screen, you will see the Welcome Back dialog box.

You will then be given the option to create a Key Disk which can be used to enable Exploration Rights on school computers running the School/Network version.

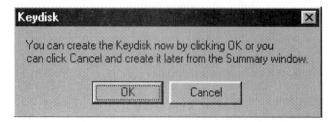

The Chapter screen is then displayed.

The Explorer Chapter Screen

In the top right corner of the Chapter screen is a Course Introduction button which allows you to play the Introduction Video or launch to the Prentice Hall website for your book.

Each Chapter in the Chapter screen is "hotlinked", taking you to the Chapter that is clicked.

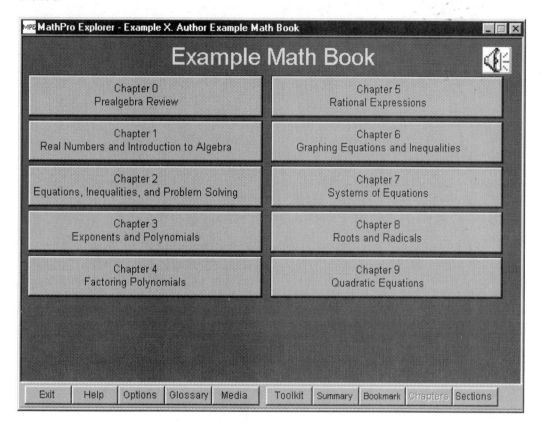

The Explorer Section Screen

The Section screen displays all the sections within a chapter. Clicking a Section button will take you to an Objective screen that shows the objectives for each section.

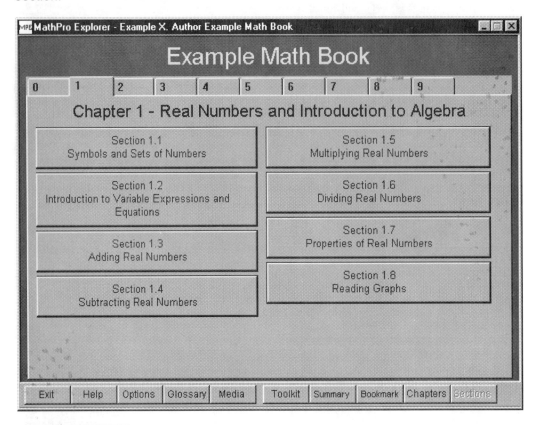

If the administrator has created a Practice Test for a particular chapter, you will see a Practice Test button to the left of the Chapter title. Clicking this button will open the Practice Test.

At the bottom of the screen is a row of buttons whereby you can:

- Access Help
- Set Options
- Access the Glossary
- View the Media Browser in which you may now also view Explorations
- Access the Toolkit
- See a Summary of your scores
- Set a Bookmark
- Go to the Chapter Screen
- Exit the program

The Practice Test Screen

The Practice Test screen displays the practice test created by your administrator/instructor. After answering a problem, click the Next Problem button or press Enter. When you have finished the test, click the Submit Test button. Before you submit your test, you will have an opportunity to go back to a problem to change your answer by clicking on the problem. In the upper right corner, the screen displays how many times you have taken this Chapter Test, as well as the maximum number of times it can be taken if constrained by the administrator.

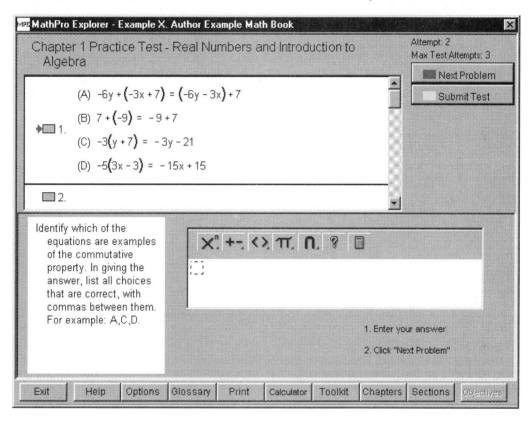

The Explorer Objective Screen

The Objective screen displays the Objective titles for each section in your book.

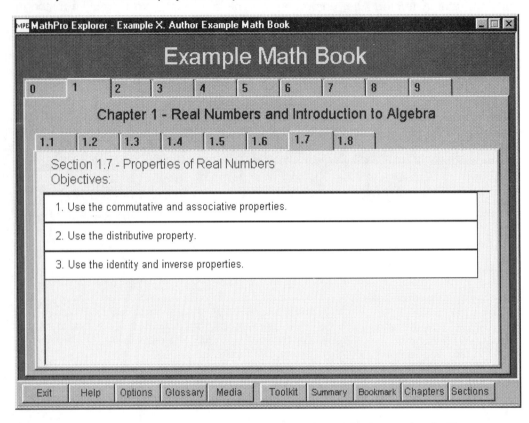

The objectives for each section correlate to those found in the textbook. To generate problems, click the Objective that you are currently studying or would like to practice. The program automatically generates 5 Warm-up problems or 10 Exercise problems each time you click unless you have specified a different number using Options.

The Explorer Media Browser

The Media button displays the Media Browser which contains a list of all explorations. You may select and open an exploration from the Browser at any time.

If your book contains Example Videos, the Browser will have an additional tab, the Video tab. You may toggle between the Explorations and the Video by clicking on a tab. Each tab contains a list of all explorations or videos. You may select and open an exploration or a video from the Browser at any time.

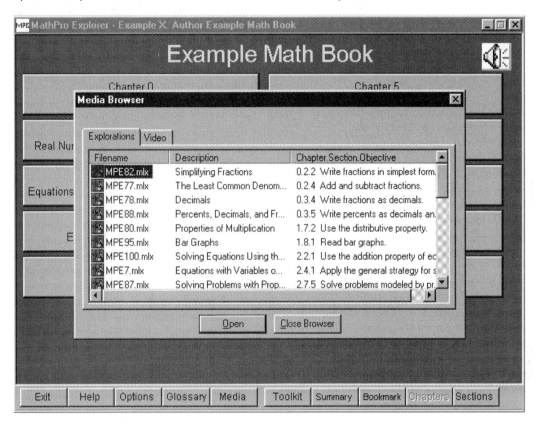

The Explorer Problem Sets Screen

When you click an objective in the Objective screen, a set of Warm-up problems or Exercise problems is automatically generated.

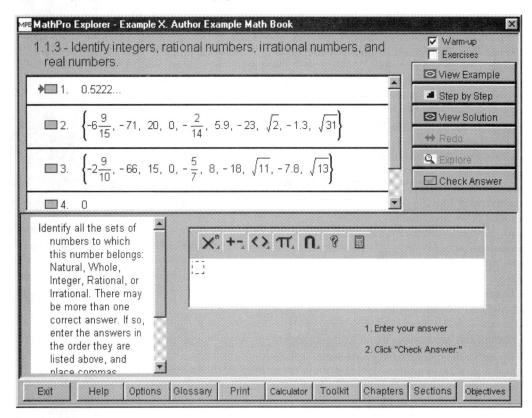

If you want Warm-up problems click, on the box to the left of Warm-ups. If you want Exercise problems, click on the box to the left of Exercises. Your scores are collected only on Exercise problems.The Print button in the toolbar at the bottom of the Problem Sets screen allows you to print out problem and status information for the current set.
Click the problem you want to work. A red arrow points to the selected problem.

Instructions are in the lower left window for each problem. These should be read before entering any solution, as these may change depending on which problem has been selected. The answer box is displayed under the set of problems. There are a variety of different displays into which results are entered. These range from a math editor, to a text box, to many different graphing displays. In some cases, there may be two boxes to enter multiple answers. Follow the instructions on the screen carefully. Read the instructions and enter your answer. Do not use a period in your answer. For example, "5 inches" can be written as "5 in" but not as "5 in."

After entering your answer, check it by clicking the Check Answer button or hitting the Enter key.

Check Answer

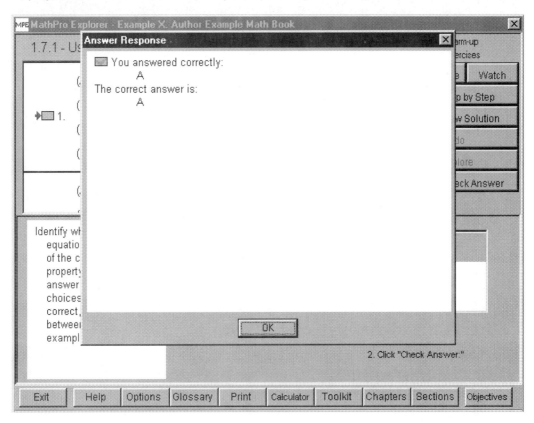 Once you have entered an answer in the given answer display, click Check Answer or press the Enter key to see if your answer is correct.

View Example or Example/Watch

[View Example] or [Example] [Watch] When you have selected a problem, click the View Example or Example button to see the solution to a problem that is similar to the one you have selected. If your book contains Example Videos you will have an additional button, the Watch button. Click the Watch button to view a video related to the current objective.

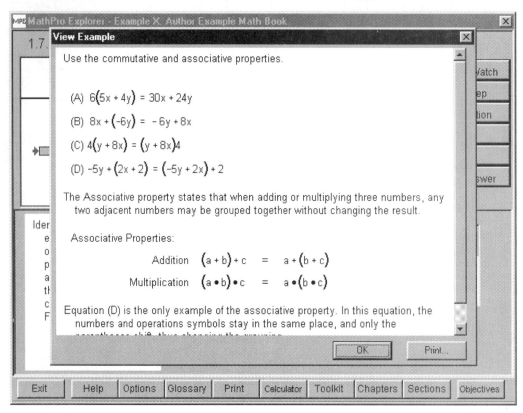

Step by Step

Step by Step   Click this button if you have difficulty solving the current problem. The Step by Step utility guides you through the process of solving each problem one step at a time. Use it to pinpoint the steps where you need help. Use it as often as you like within an objective until you have mastered the process for solving this kind of problem. You may only use Step by Step once per problem. If you use the Step by Step utility, you will not receive credit for this problem.

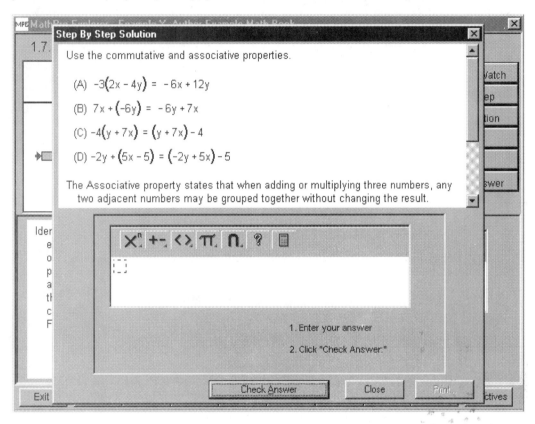

Step By Step Solution

Use the commutative and associative properties.

(A) $-3(2x - 4y) = -6x + 12y$

(B) $7x + (-6y) = -6y + 7x$

(C) $-4(y + 7x) = (y + 7x) - 4$

(D) $-2y + (5x - 5) = (-2y + 5x) - 5$

The Associative property states that when adding or multiplying three numbers, any two adjacent numbers may be grouped together without changing the result.

1. Enter your answer

2. Click "Check Answer:"

Check Answer   Close   Print

21

View Solution

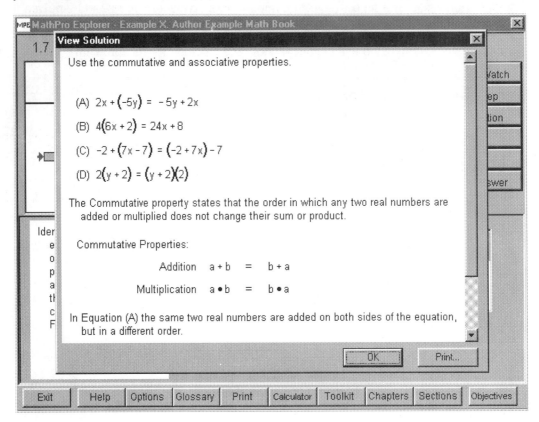 Click this button to view the solution to the current problem. If you use the View Solution utility, you will not receive credit for this problem.

View Solution

Use the commutative and associative properties.

(A) $2x + (-5y) = -5y + 2x$

(B) $4(6x + 2) = 24x + 8$

(C) $-2 + (7x - 7) = (-2 + 7x) - 7$

(D) $2(y + 2) = (y + 2)(2)$

The Commutative property states that the order in which any two real numbers are added or multiplied does not change their sum or product.

Commutative Properties:

Addition $\quad a + b \quad = \quad b + a$

Multiplication $\quad a \bullet b \quad = \quad b \bullet a$

In Equation (A) the same two real numbers are added on both sides of the equation, but in a different order.

OK | Print...

Exit | Help | Options | Glossary | Print | Calculator | Toolkit | Chapters | Sections | Objectives

Redo

Redo If you click Check Answer and the problem is incorrect, you have the option to go back and work the same type of problem. Simply click on the incorrectly answered problem and click the Redo button. A new problem is generated. The incorrect problem and the new problem are tracked and the scores are added to the cumulative scores for the objective if you are working Exercise problems.

22

Explore

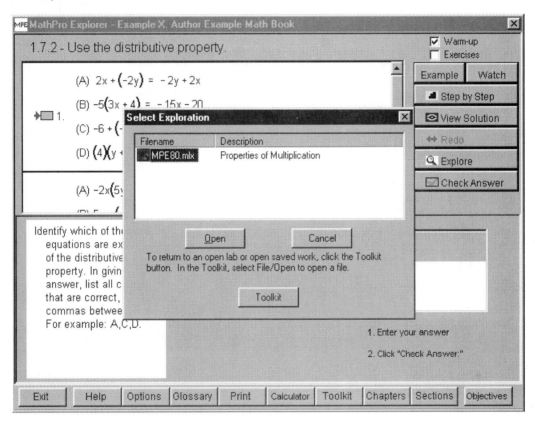

The Explore button is displayed in one of two conditions. If there is an Exploration associated with the current objective, the Explore button is active. By clicking the active button, you may open an Exploration associated with this objective.

If there are no Explorations associated with the objective, the Explore button is inactive. If you click on the inactive Explore button you will see the following dialog box:

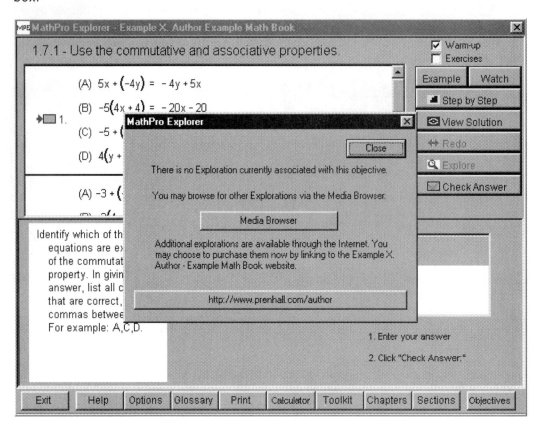

You may select the Media Browser to see a list of all existing Explorations for your book.

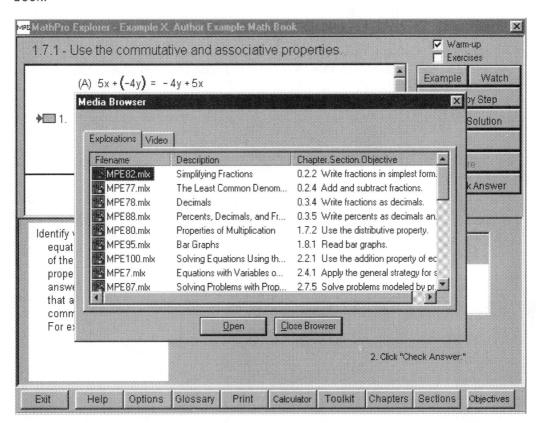

You may also launch the Media Browser at any time from the Media button on the main toolbar from the Chapter, Section or Objectives screens.

The **Chapters, Sections** and **Objectives** buttons will close the problem window and return you to the selected location.

MathPro Explorer Toolkit

You may open Explorations from the Explore button, the Media Browser, or you may click the Toolkit button on the main toolbar to access the Toolkit. If you have upgraded to Explorations Rights, you may use the Toolkit to work any existing Explorations or open your previously saved files. If you have upgraded to the Full Toolkit Rights you have access to all tools in the Toolkit and may create new labs and work independently.

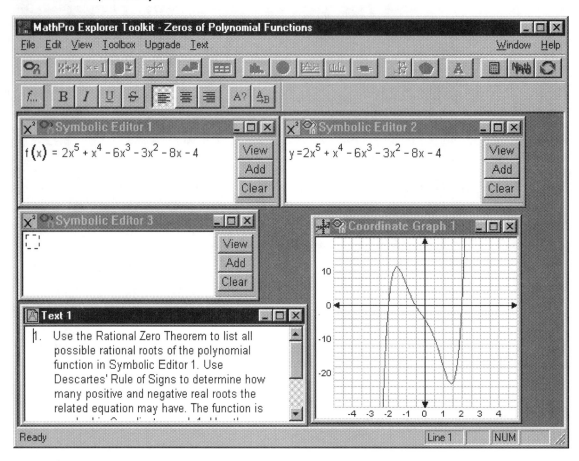

Explorer Button Definitions

Help

Help   Click this button at any time to get Help.

Options

Options   Click Options to change basic MathPro Explorer settings.

Enable Quick Help - when checked, this will bring up small help boxes when the cursor is placed over an active area on the screen.

Number of Problems Generated - Change the numbers to match the number of Warm-up and Exercise problems you want to generate.

Glossary

Glossary   Use the Glossary to study the terminology used in the course or to look up a specific word.

Media

Media   Click the Media button to open the Media Browser which lists all Explorations for this book. You may select and open an Exploration from the Browser at any time.

If your book contains Example Videos, the Media Browser lists all Explorations and Example Videos for this book. You may select and open an Exploration or Video from the Browser at any time.

Toolkit

Toolkit   Click the Toolkit Button to access the Toolkit. If you have upgraded to Explorations Rights, you may use the Toolkit to work any existing Explorations or open your previously saved files. If you have upgraded to the Full Toolkit Rights you have access to all tools in the Toolkit and may create new labs and work independently.

Calculator

Calculator   Click the Calculator button to bring up the Windows Calculator. This is helpful with many of the problems that require large calculations. You can switch between the standard and the scientific calculator by clicking View in the menu bar and choosing the calculator you want.

Summary

Summary You can view your performance by Section, Chapter, Date, or Book. Tabs allow you to select the level of information. A drop down list in each tab enables you to select scores for different sections. Your scores are collected only on Exercise problems. If you have generated more than one set of Exercise problems, the scores from each set are saved.

Bookmark

Bookmark Use Bookmark when you see particular sections that you would like to revisit. Also, if you are interrupted while using MathPro Explorer, Bookmark will save your place. Click Set in the section to which you want to return to set a bookmark. Click Remove to delete the bookmark from a selected section. When you want to return to a bookmarked section, just reopen Bookmark and click the desired section.

Chapters

Chapters Click Chapters to bring up the Chapter screen which displays all the chapters in the book.

Sections

Sections Click Sections to bring up the Sections screen within a chapter.

Exit

Exit Leave MathPro Explorer at any time by clicking Exit. You will be asked if you are sure you want to quit before the program exits.

Print

Print Click the Print button in the toolbar at the bottom of the Problem Sets to print out problem and status information for the current set.

Watch

Watch If your book contains Example Videos, click the Watch button to view a video of the author working a problem related to the current objective.

Check Answer

Check Answer Once you have entered an answer in the given answer display, click Check Answer or press the Enter key to see if your answer is correct.

View Example/Example

View Example or **Example** When you have selected a problem, click the View Example or Example button to see the solution to a problem that is similar to the one you have selected. Use it as often as you like until you have mastered the process for solving this kind of problem.

Step by Step

Step by Step Click this button if you have difficulty solving the current problem. The Step by Step utility guides you through the process of solving each problem one step at a time. Use it to pinpoint the steps where you need help. Use it as often as you like within an objective until you have mastered the process for solving this kind of problem. You may only use Step by Step once per problem. If you use the Step by Step utility, you will not receive credit for this problem.

View Solution

View Solution Click this to view the solution to the current problem. If you use the View Solution utility, you will not receive credit for this problem.

Redo

Redo If you click Check Answer and the problem is incorrect, you have the option to go back and work the same type of problem. Simply click on the incorrectly answered problem and click the Redo button. A new problem is generated. The incorrect problem and the new problem are tracked and the scores are added to the cumulative scores for the objective if you are working Exercise problems.

Explore

Explore The Explore button is displayed in one of two conditions. If there is an Exploration associated with the current objective, the Explore button is active. By clicking the active button, you may open an Exploration associated with this objective. If there are no Explorations associated with the objective, the Explore button is inactive. If you click on the inactive Explore button you will see a dialog box from which you may select the Media Browser to see a list of all existing Explorations for your book or you may launch the Prentice Hall website for your book.

Symbol Buttons

The small red arrow in the lower corner of the buttons indicates that there is a flip-up menu with additional buttons. Click any button to access its sub-menu. When a flip-up menu is showing, you can use the following for keyboard navigation:

Arrows - move up, down, right, left
Space bar - selects the cell you have focus on
Enter - functions the same as the space bar
Esc - closes the flip-up menu

Expressions - Click this button to access exponent, subscript, square root, nth root, rational, and absolute value symbols.

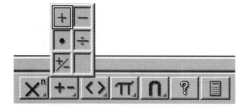

Binary Operators - Click this button to access addition, subtraction, multiplication, division and plus/minus symbols.

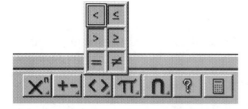

Relational Operators - Click this button to access less than, less than or equal to, greater than, greater than or equal to, equal or not equal symbols.

Constants - Click this button to access constant e, constant i, pi, or infinity symbols.

Set Operators - Click this button to access the set union, set intersection, member of set and not member of set symbols.

Expressions

This Icon will enable you to enter a rational expression.

1) To enter the fraction $\dfrac{2}{3}$, click the Icon, then type 2 in the top box, press the right arrow key, or click in the bottom box, and type 3.

2) If you have already typed an expression and want to put it in the numerator of a rational, highlight the expression and click the button. For example, if you typed the expression $3x^3+2y$ and you want to put it in the numerator, highlight the expression $3x^3+2y$ and click the button. This will give you the expression $\dfrac{3x^3+2y}{\square}$.

This Icon will enable you to enter an exponent.

1) To enter x^2, type x and click the Icon. This gives you the expression $x^{\square}$. Type a 2, this will appear in the exponent box.

2) If you have already typed an expression and want to put it in an exponent, highlight the expression and click the button. For example, if you typed the expression $3x^3+2y$ and you want to put it in an exponent, highlight the expression $3x^3+2y$ and click the button. This will give you the expression $\square^{3x^3+2y}$. You can click in the box or hit the left arrow key to enter a base.

This Icon will enable you to enter subscripts.

1) To enter x_1, type x and click the 🔲 Icon. This will give you $x_{\square}$, type in a 1 in the subscript box.

2) If you have already typed an expression and want to put it in a subscript, highlight the expression and click the 🔲 button. For example, if you typed the expression $3x^3+2y$ and you want to put it in a subscript, highlight the expression $3x^3+2y$ and click the 🔲 button. This will give you the expression $\square_{3x^3+2y}$. You can click in the box or hit the left arrow key to enter a base.

🔲 This Icon will enable you to enter a square root.

1) To enter $\sqrt{x+9}$ click the 🔲 and type x + 9 in the radicand box.

2) If you have already entered an expression and want to put it in a radical, highlight the expression and click the 🔲 button. For example if you typed

$$\frac{-b\pm b^2-4ac}{2a}$$

and you wanted to put the expression b^2-4ac under a radical, highlight just that part of the expression and click the 🔲 button. This will give you the expression

$$\frac{-b\pm\sqrt{b^2-4ac}}{2a}.$$

🔲 This icon will enable you to enter any root.

1) To enter $\sqrt[4]{16}$, click the 🔲 symbol, the cursor will be in the smaller box outside the radical. Type 4, then click in the box under the radical or press the right arrow key, and type 16.

2) If you have already typed an expression, you can highlight the expression and click the 🔲 button. This will put the expression under the radical and put the cursor in the small box outside the radical. For example if you typed the expression $3x^3+2y$ and wanted the cube root of it, highlight the expression $3x^3+2y$ and click the 🔲 button. This will give you $\sqrt[\square]{3x^3+2y}$. The cursor is already in the box outside the radical, so type 3 to get the cube root expression $\sqrt[3]{3x^3+2y}$.

⊠ This icon puts up absolute value bars with a box inside in which you can enter your answer.

Binary Operators

+   Inserts a plus sign at the position of the cursor.

−   Inserts a minus sign at the position of the cursor.

·   Inserts a multiplication sign at the position of the cursor.

÷   Inserts a division sign at the position of the cursor.

±   Inserts a plus/minus sign at the position of the cursor.

Relational Operators

<   Inserts a less than symbol at the position of the cursor.

≤   Inserts a less than or equal to symbol at the position of the cursor.

>   Inserts a greater than symbol at the position of the cursor.

≥   Inserts a greater than or equal to symbol at the position of the cursor.

=   Inserts a equal to symbol at the position of the cursor.

≠   Inserts a not equal symbol at the position of the cursor.

Constants

e Inserts a constant e symbol at the position of the cursor.

i Inserts a constant i symbol at the position of the cursor.

π Inserts a Pi symbol at the position of the cursor.

∞   Inserts an infinity symbol at the position of the cursor.

Set Operators

∪ Inserts a set union symbol at the position of the cursor.

∩ Inserts a set intersection symbol at the position of the cursor.

∈ Inserts a member of set symbol at the position of the cursor.

∉ Inserts a not member of set symbol at the position of the cursor.

Special Buttons

? Click the Help icon button to open the Entering Math Expressions screen of the Help file.

▦ Click the Calculator button to bring up the Windows Calculator. This is helpful with many of the problems that require large calculations. You can switch between the standard and the scientific calculator by clicking View in the menu bar and choosing the calculator you want.

Entering Math Expressions

You can enter math symbols using the keyboard or the buttons on the Math Editor toolbar submenus.

Two Methods for Using the Buttons

1) Click the button to access the submenu for the desired expression. Select the specific symbol needed. That expression will be inserted in the Math Editor Window at the position of the blinking cursor. Use the mouse to move the cursor to the desired dotted rectangle then type your answer.

   Use the mouse or the right and left arrow keys to move the position of the cursor.

2) If there is red on a button in a pop-up submenu, you can highlight an expression in the Math Editor Window and click the desired button of the symbol you want. This will insert that expression where you see the red in the Icon box.

Using the Keyboard
*Ctrl- indicates you must hold down the <Ctrl> key while pressing the following key.

Special/Editing Keys
Ctrl-Z: Undo the most recent operation or Alt-Backspace
Ctrl-X: Cuts a section out of the expression
Ctrl-C: Copies a section
Ctrl-V: Pastes the most recent cut or copy block
Ctrl-Y: Redo the most recent operation or Shift-Alt-Backspace
Home: Start of equation
Left arrow: Left
Right arrow: Right
End: End of equation

Special Symbols Using the Keyboard:

Operation symbols:
Less Than Equal: Ctrl-,
Greater Than Equal: Ctrl-.
Not Equal: Ctrl-3
Pi: Ctrl-p
Plus-Minus: Ctrl-=
Divide: Ctrl-:
Multiply: Ctrl-8

Special structures:
Rational: Ctrl-/
Square Root: Ctrl-s
Root: Ctrl-r
Exponent: Ctrl-6
Subscript: Ctrl--
Absolute: Ctrl-A
Summation: Ctrl-U
mxn Matrix: Ctrl-M
2x2 Matrix: Ctrl-2
Vector: Ctrl-1
e: Ctrl-E
I: Ctrl-I

Log: Ctrl-L
Ln: Ctrl-N
Log Base N: Ctrl-B
Inverse Log: Ctrl-Alt-L
Inverse Ln: Ctrl-Alt-N
Sin: Ctrl-H
Cos: Ctrl-O
Tan: Ctrl-T
Inverse Sin: Ctrl-Alt-H
Inverse Cos: Ctrl-Alt-O
Inverse Tan: Ctrl-Alt-T
Permutation: Ctrl-Alt-P
Combination: Ctrl-Alt-C
Union: Ctrl-9
Intersection: Ctrl-0

Selection keys:
Select to start of equation: Shift-Home
Select char to left: Shift-Left arrow
Select char to right: Shift-Right arrow
Select to end of equation: Shift-End

Math Editor Special Keys

Special/Editing Keys
Ctrl-Z: Undo the most recent operation or Alt-Backspace
Ctrl-X: Cuts a section out of the expression
Ctrl-C: Copies a section
Ctrl-V: Pastes the most recent cut or copy block
Ctrl-Y: Redo the most recent operation or Shift-Alt-Backspace
Home: Start of equation
Left arrow: Left
Right arrow: Right
End: End of equation

Special symbols not on keyboard:
Less Than Equal: Ctrl-,
Greater Than Equal: Ctrl-.
Not Equal: Ctrl-3
Pi: Ctrl-p
Plus-Minus: Ctrl-=
Divide: Ctrl-:
Multiply: Ctrl-8

Special structures:
Rational: Ctrl-/
Square Root: Ctrl-s
Root: Ctrl-r
Exponent: Ctrl-6

Subscript: Ctrl--
Absolute: Ctrl-A
Summation: Ctrl-U
mxn Matrix: Ctrl-M
2x2 Matrix: Ctrl-2
Vector: Ctrl-1
Constant e: Ctrl-E
Constant I: Ctrl-I
Log: Ctrl-L
Ln: Ctrl-N
Log Base N: Ctrl-B
Inverse Log: Ctrl-Alt-L
Inverse Ln: Ctrl-Alt-N
Sin: Ctrl-H
Cos: Ctrl-O
Tan: Ctrl-T
Inverse Sin: Ctrl-Alt-H
Inverse Cos: Ctrl-Alt-O
Inverse Tan: Ctrl-Alt-T
Permutation: Ctrl-Alt-P
Combination: Ctrl-Alt-C
Union: Ctrl-9
Intersection: Ctrl-0

Selection keys:
Select to start of equation: Shift-Home
Select char to left: Shift-Left arrow
Select char to right: Shift-Right arrow
Select to end of equation: Shift-End

About MathPro Explorer

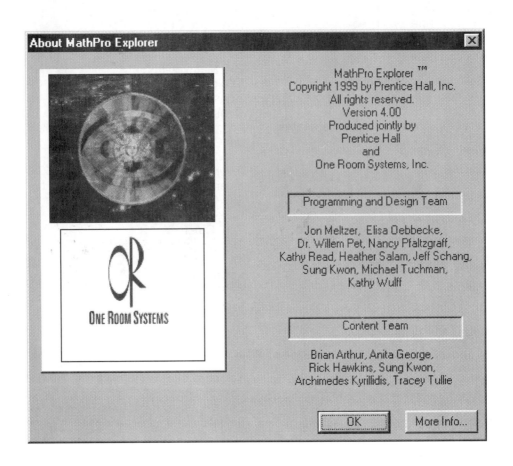

Multimedia MathPro Explorer 4.0 Home/Student Administrator Program

MathPro Explorer Administrator Overview

The MathPro Explorer Administrator Program provides you with several different options. You may generate reports or retrieve saved reports, which include results from Exercise problems and Practice Tests. In addition, you may create Practice Tests to take in MathPro Explorer. The On-Line Help feature is available for convenient assistance.

Administrator Login

Upon entering the MathPro Explorer Administrator Program, you are prompted to enter a password. The default is no password; therefore, the password should be set the first time you enter the Administrator Program. Click OK on the Login screen to open the MathPro Explorer Administrator main menu.

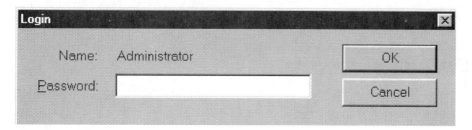

MathPro Explorer Administrator Main Menu

The MathPro Explorer Administrator main menu is displayed below. You have the option to use the main menu or the toolbar buttons to navigate through the program.

Toolbar Buttons

The New Report button allows you to generate a Summary Report or Practice Test Report.

The Open Report button allows you to open a previously saved Summary Report or Practice Test Report.

The Save Report button allows you to save a current Summary Report or Practice Test Report.

The Print button allows you to print an open Summary Report or Practice Test Report.

The Refresh button allows you to refresh an open Summary Report if you are working Exercise problems in MathPro Explorer.

The Help button allows you to access the on-line Administrator help menu at anytime.

Creating Reports

To generate a report, select File from the main menu, then select New Report as displayed below or click the New Report button.

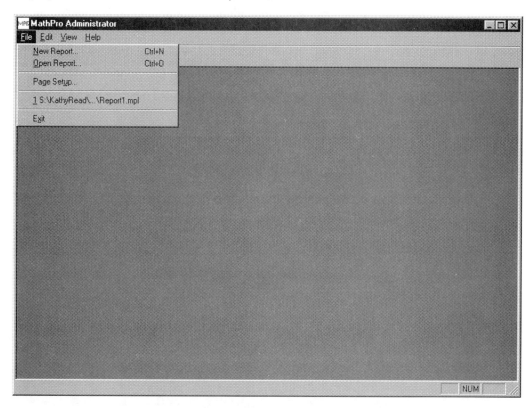

Once New Report is selected, the following dialog box opens.

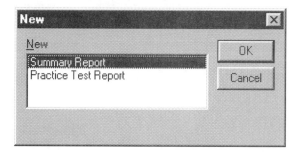

Select the type of report to generate, Summary Report or Practice Test Report, and click OK.

Summary Report

Type the desired Chapter information in the corresponding field or use the list box to select the desired Chapter information for the report. Once this information is entered, use your mouse to check the desired data to include (Book, Chapter, Section, Objective, Cumulative, or By Date) and click the View Report button.

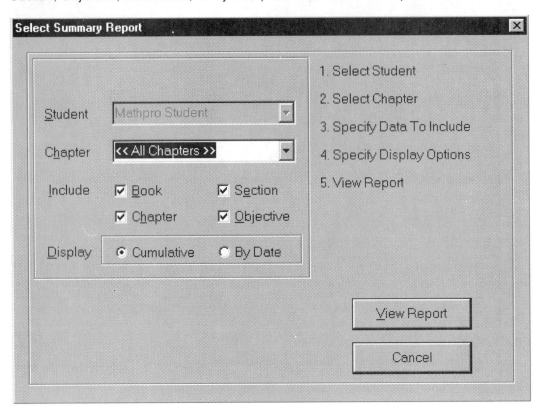

Once the View Report button is selected, a report is generated similar to the one shown below. (The generated report is a text file that can be viewed, saved and reopened, or printed.)

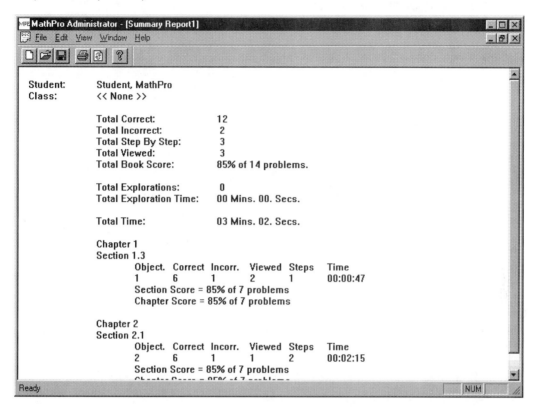

Practice Test Report

Type the desired Chapter information in the corresponding field or use the list box to select the desired Chapter information for the report. Once this information is entered, click the View Report button.

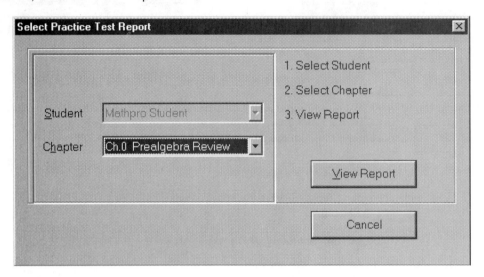

Once the View Report button is selected, a report is generated similar to the one shown below. (The generated report is a text file that can be viewed, saved and reopened, or printed.)

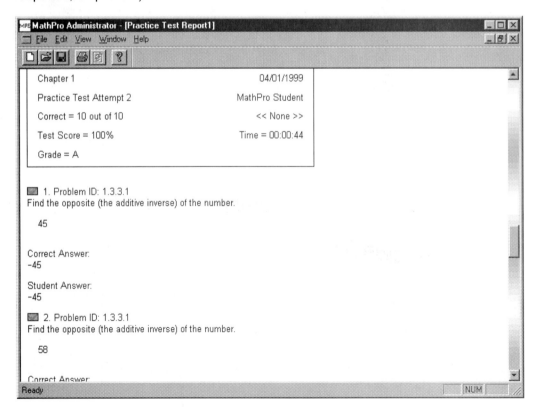

Creating Practice Tests and Changing the Password

From the MathPro Explorer Administrator Program, you have the option to create Practice Tests or Change Password. These options may be selected from the Edit menu.

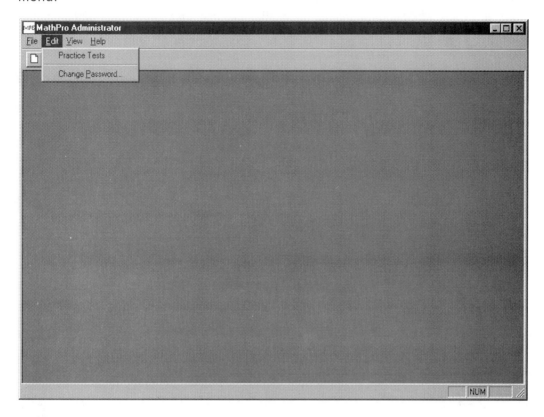

Creating Practice Tests

Select Practice Tests from the Edit menu to open the Practice Chapter Tests dialog box. Through this screen, you may enter the chapter, the number of problems, the maximum number of tries and the minimum grade percentages desired to create a test.

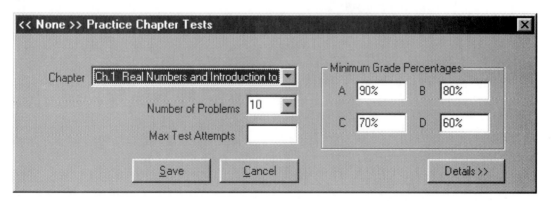

Click the Details button to view the chapter problems. Use the mouse to navigate through the tree structure on the left side of the dialog box under Chapter Problems, highlight a problem, and view a Sample Problem in the lower portion of the screen. The Fill button creates a randomly generated set of problems from the chapter dependent upon the Number of Problems selected. The Add button adds a highlighted problem to the Selected Problem list on the right side of the dialog box. Once problems have been selected, you have the option to remove one problem or remove all problems by clicking on the Remove or Remove All buttons, respectively.

<< None >> Practice Chapter Tests ☒

Chapter [Ch.1 Real Numbers and Introduction to ▼]

┌ Minimum Grade Percentages ──────────┐
│ A [90%]   B [80%]   │
│ C [70%]   D [60%]   │
└──────────────────────────────────────┘

Number of Problems [10 ▼]

Max Test Attempts [   ]

[Save]   [Cancel]   [Details <<]

Chapter Problems

```
⊟ Obj.2 Evaluate algebraic exp ▲
    [1.2.2.1]
    1.2.2.2
    1.2.2.3
    1.2.2.4          ▼
◄                    ►
```

[Fill]

[Add >>]

[<< Remove]

[<< Remove All]

Selected Problems

```
1.2.1.4
1.2.1.15
1.2.2.1
```

Sample Problem

1.2.2.1
Evaluate.

$$8x + 1 \quad \text{for} \quad x = \frac{3}{8}$$

Changing the Password

Select Change Password from the Edit menu to open the Change Password dialog box. Type a new password in the New Password field and repeat it in the Confirm Password field. Click the OK button to initiate the change or click Cancel to abort the change. Once the password is changed, remember it to log back into the MathPro Explorer Administrator Program.

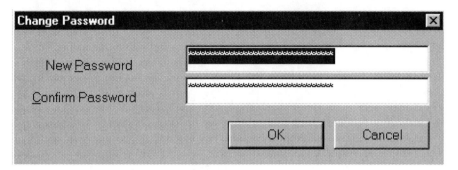

Record Keeping Information

Options

Click the Options button to set defaults within the program.

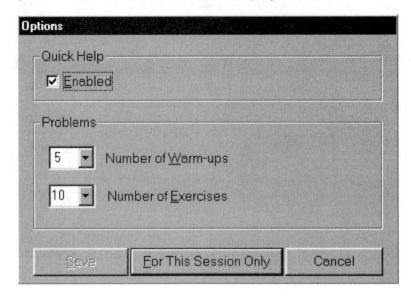

Check or uncheck Enabled to turn the Quick Help tool tip information boxes on or off. The list boxes in the Problems section are used to set the number of Warm-up problems and/or Exercise problems to generate and track in each objective. These settings can be saved as the default or saved only for this working session.

Section Records

Student information is collected only on Exercise problems. After Exercise problems have been worked in an objective, a record of your activities is recorded.

Click the Summary button from the Chapter, Section, or Objective screens to open the Scores For <User> dialog box. The Section Scores information gives you feedback on the work completed in each section. The information includes:

- A summary of all objectives worked within a section
- The number of Exercise problems solved correctly or incorrectly for each objective
- The number of times View Solution and Step by Step have been accessed
- The length of time spent on Exercises for each objective
- The names of any Explorations worked for each objective
- The amount of time spent on each Exploration
- A cumulative total for each of these categories, as well as a cumulative score percentage for the section.

Click the Print button to print a report which includes Section Score information.

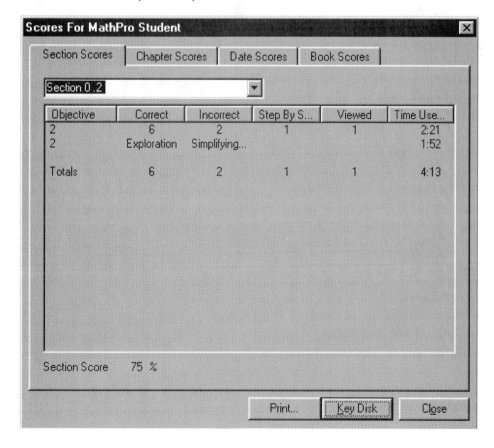

Chapter Records

The Chapter Scores information provides:

- A summary of performance by chapter
- A summary of all sections worked within each chapter
- The number of Exercise problems solved correctly or incorrectly for each section
- The number of times View Solution and Step-by-Step have been accessed
- The number of Explorations worked for each section
- The amount of time spent on each section
- A cumulative total for each of these categories, as well as a cumulative score percentage for the chapter.

Click the Print button to print a report which includes Chapter Score information.

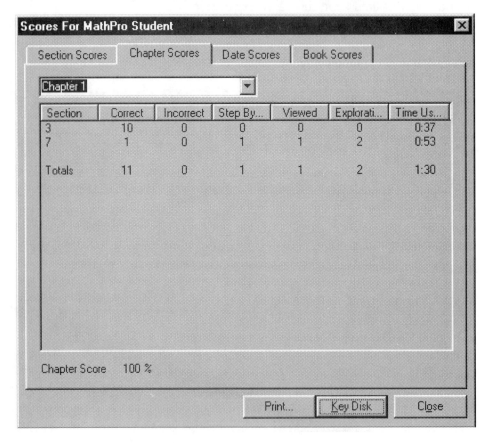

Date Records

The Date Scores information provides:

- A summary of all objectives worked within a section on a specific date
- The number of Exercise problems solved correctly or incorrectly for each objective
- The number of times View Solution and Step by Step have been accessed
- The length of time spent on Exercises for each objective
- The names of any Explorations worked for each objective
- The amount of time spent on each Exploration
- A cumulative total for each of these categories, as well as a cumulative score percentage for the section.

Click the Print button to print a report which includes Date Score information.

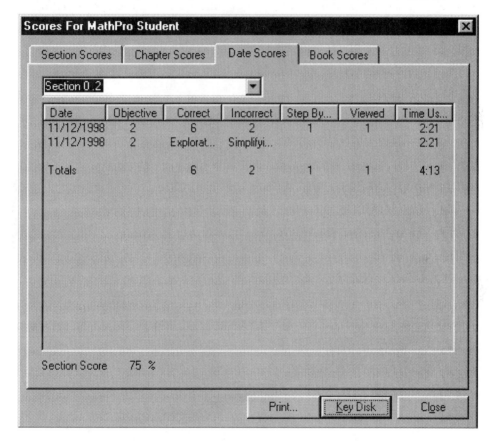

Book Records

The Book Scores information provides:

- The book totals of Exercise problems solved correctly and incorrectly
- The number of times View Solution and Step by Step have been accessed
- The number of Explorations worked
- The total amount of time spent on Explorations
- The total number of Exercise problems attempted in the book
- The percentage of correct answers
- The amount of time spent on Exercise problems
- The number of sessions and the total time spent working Exercise problems in MathPro Explorer.

Click the Print button to print a report which includes Book Score information.

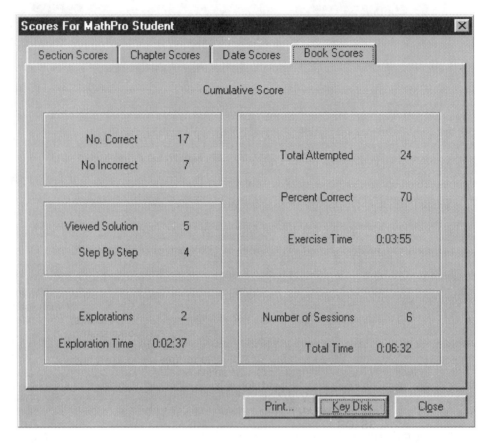

SINGLE PC LICENSE AGREEMENT AND LIMITED WARRANTY

READ THIS LICENSE CAREFULLY BEFORE OPENING THIS PACKAGE. BY OPENING THIS PACKAGE, YOU ARE AGREEING TO THE TERMS AND CONDITIONS OF THIS LICENSE. IF YOU DO NOT AGREE, DO NOT OPEN THE PACKAGE. PROMPTLY RETURN THE UNOPENED PACKAGE AND ALL ACCOMPANYING ITEMS TO THE PLACE YOU OBTAINED THEM [[FOR A FULL REFUND OF ANY SUMS YOU HAVE PAID FOR THE SOFTWARE]]. THESE TERMS APPLY TO ALL LICENSED SOFTWARE ON THE DISK EXCEPT THAT THE TERMS FOR USE OF ANY SHAREWARE OR FREEWARE ON THE DISKETTES ARE AS SET FORTH IN THE ELECTRONIC LICENSE LOCATED ON THE DISK:

1. GRANT OF LICENSE and OWNERSHIP: The enclosed computer programs and data ("Software") are licensed, not sold, to you by Prentice-Hall, Inc. ("We" or the "Company") and in consideration of your purchase or adoption of the accompanying Company textbooks and/or other materials, and your agreement to these terms. We reserve any rights not granted to you. You own only the disk(s) but we and/or our licensors own the Software itself. This license allows you to use and display your copy of the Software on a single computer (i.e., with a single CPU) at a single location for academic use only, so long as you comply with the terms of this Agreement. You may make one copy for back up, or transfer your copy to another CPU, provided that the Software is usable on only one computer.

2. RESTRICTIONS: You may not transfer or distribute the Software or documentation to anyone else. Except for backup, you may not copy the documentation or the Software. You may not network the Software or otherwise use it on more than one computer or computer terminal at the same time. You may not reverse engineer, disassemble, decompile, modify, adapt, translate, or create derivative works based on the Software or the Documentation. You may be held legally responsible for any copying or copyright infringement which is caused by your failure to abide by the terms of these restrictions.

3. TERMINATION: This license is effective until terminated. This license will terminate automatically without notice from the Company if you fail to comply with any provisions or limitations of this license. Upon termination, you shall destroy the Documentation and all copies of the Software. All provisions of this Agreement as to limitation and disclaimer of warranties, limitation of liability, remedies or damages, and our ownership rights shall survive termination.

4. LIMITED WARRANTY AND DISCLAIMER OF WARRANTY: Company warrants that for a period of 60 days from the date you purchase this SOFTWARE (or purchase or adopt the accompanying textbook), the Software, when properly installed and used in accordance with the Documentation, will operate in substantial conformity with the description of the Software set forth in the Documentation, and that for a period of 30 days the disk(s) on which the Software is delivered shall be free from defects in materials and workmanship under normal use. The Company does not warrant that the Software will meet your requirements or that the operation of the Software will be uninterrupted or error-free. Your only remedy and the Company's only obligation under these limited warranties is, at the Company's option, return of the disk for a refund of any amounts paid for it by you or replacement of the disk. THIS LIMITED WARRANTY IS THE ONLY WARRANTY PROVIDED BY THE COMPANY AND ITS LICENSORS, AND THE COMPANY AND ITS LICENSORS DISCLAIM ALL OTHER WARRANTIES, EXPRESS OR IMPLIED, INCLUDING WITHOUT LIMITATION, THE IMPLIED WARRANTIES OF MERCHANTABILITY AND FITNESS FOR A PARTICULAR PURPOSE. THE COMPANY DOES NOT WARRANT, GUARANTEE OR MAKE ANY REPRESENTATION REGARDING THE ACCURACY, RELIABILITY, CURRENTNESS, USE, OR RESULTS OF USE, OF THE SOFTWARE.

5. LIMITATION OF REMEDIES AND DAMAGES: IN NO EVENT, SHALL THE COMPANY OR ITS EMPLOYEES, AGENTS, LICENSORS, OR CONTRACTORS BE LIABLE FOR ANY INCIDENTAL, INDIRECT, SPECIAL, OR CONSEQUENTIAL DAMAGES ARISING OUT OF OR IN CONNECTION WITH THIS LICENSE OR THE SOFTWARE, INCLUDING FOR LOSS OF USE, LOSS OF DATA, LOSS OF INCOME OR PROFIT, OR OTHER LOSSES, SUSTAINED AS A RESULT OF INJURY TO ANY PERSON, OR LOSS OF OR DAMAGE TO PROPERTY, OR CLAIMS OF THIRD PARTIES, EVEN IF THE COMPANY OR AN AUTHORIZED REPRESENTATIVE OF THE COMPANY HAS BEEN ADVISED OF THE POSSIBILITY OF

SUCH DAMAGES. IN NO EVENT SHALL THE LIABILITY OF THE COMPANY FOR DAMAGES WITH RESPECT TO THE SOFTWARE EXCEED THE AMOUNTS ACTUALLY PAID BY YOU, IF ANY, FOR THE SOFTWARE OR THE ACCOMPANYING TEXTBOOK. BECAUSE SOME JURISDICTIONS DO NOT ALLOW THE LIMITATION OF LIABILITY IN CERTAIN CIRCUMSTANCES, THE ABOVE LIMITATIONS MAY NOT ALWAYS APPLY TO YOU.

6. GENERAL: THIS AGREEMENT SHALL BE CONSTRUED IN ACCORDANCE WITH THE LAWS OF THE UNITED STATES OF AMERICA AND THE STATE OF NEW YORK, APPLICABLE TO CONTRACTS MADE IN NEW YORK, AND SHALL BENEFIT THE COMPANY, ITS AFFILIATES AND ASSIGNEES. HIS AGREEMENT IS THE COMPLETE AND EXCLUSIVE STATEMENT OF THE AGREEMENT BETWEEN YOU AND THE COMPANY AND SUPERSEDES ALL PROPOSALS OR PRIOR AGREEMENTS, ORAL, OR WRITTEN, AND ANY OTHER COMMUNICATIONS BETWEEN YOU AND THE COMPANY OR ANY REPRESENTATIVE OF THE COMPANY RELATING TO THE SUBJECT MATTER OF THIS AGREEMENT. If you are a U.S. Government user, this Software is licensed with "restricted rights" as set forth in subparagraphs (a)-(d) of the Commercial Computer-Restricted Rights clause at FAR 52.227-19 or in subparagraphs (c)(1)(ii) of the Rights in Technical Data and Computer Software clause at DFARS 252.227-7013, and similar clauses, as applicable.

Should you have any questions concerning this agreement or if you wish to contact the Company for any reason, please contact in writing: New Media / Higher Education Division / Prentice Hall Inc. / 1 Lake Street / Upper Saddle River, NJ 07458.